Amazon Echo Dot

The Ultimate Amazon Echo User Guide!

Table of Contents

iv

Introduction

Thank you for downloading Amazon Echo Dot: The Ultimate Amazon Echo User Guide!

Amazon Echo Dot is not just a speaker – this is your new virtual assistant, designed to do everything that you need to get done. This book is designed to help you get to master using your Amazon Echo Dot and turn it into the best virtual assistant ever.

This book will serve as your all-in-one guide to Amazon Echo Dot (2nd Generation), the newest and hottest Amazon product out in the market. In this book, you will learn how to do the following:

- Set up your Echo Dot for first-time use
- Connect your Echo Dot to your other apps
- Use it as a remote control for your smart home
- Discover other tasks that Alexa can do
- Make your Amazon Echo Dot adapt to your needs
- Troubleshoot your Amazon Echo Dot

and so much more!

If you are ready to begin making every room in your house awesome with Amazon Echo Dot, jump in on the first chapter!

Chapter 1: Why Choose Amazon Echo Dot?

If you owned an Amazon Echo, or at least have heard of them, you are probably aware that it is more than just a speaker – it is a virtual assistant that will make you feel that you are Tony Stark and you have your very own Jarvis in your home.

Amazon Echo does not have a screen – it only has a speaker, which is designed to pick up your voice commands and return with an audio response. Alexa, the software that runs Amazon Echo, will serve as your cloud-based, voice-controlled personal assistant.

If you have long wanted to have an Amazon Echo, but you decided against buying it because of the price, here is some good news: Amazon has released the Amazon Echo Dot, which is the cheaper and smaller version of Amazon Echo.

What is Amazon Echo Dot?

The Echo Dot is a device that uses the software Alexa in order to issue voice-controlled and hands-free commands, such as setting up your daily alarms, read you the news, control devices in your smart home, provide information found on the Internet, order food, and that's just for starters!

The device can "hear" you across the room thanks to its built-in seven far-field microphones which are designed to pick up your voice even while you are playing music or in a noisy environment. It also has a built-in speaker that will allow you to use it anywhere that would require a voice-activated computer or an alarm clock.

The feature that truly makes the Echo Dot amazing is that it is powered by the extremely smart software named Alexa, which takes care of all the commands that you issue. You can think of Alexa as similar to Siri, but this assistant is capable of learning more skills and adapt to your vocabulary, speech patterns, and personal preferences. Since your Echo Dot is always connected to the internet, you can expect Alexa always to be updated, as new skills are rolled out now and then.

Amazon Echo Dot is available on Amazon.com at about $50, which is about 1/3 of the Amazon Echo price, which is about $140. For this reason, some consumers opt to get multiple Dots to have an Alexa in all of their rooms. Take note that if you have two or more Dots in a room, the device that will respond to you is the one closest to you, thanks to Echo's feature called Echo Spatial Perception.

Unboxing and Feature Overview

What comes with the box is a 9w power adapter, a quick start guide, and a micro-USB cable for charging the Echo Dot. You can also purchase compatible accessories that you may need shortly, such as the 3.5mm audio cable, the Echo Dot case (comes in different colors), and the Alexa voice remote (necessary if you want to extend the range wherein you and Alexa can hear each other).

If you want to use it right away, all you need to do is to plug in the device and connect the device to the internet using the Alexa app. Once it gets hooked on the internet, Alexa will greet you and prompt that it is ready to receive your commands.

What makes the Echo Dot special is that it has very simple and intuitive features – all you need to press in this device are four buttons – the up and down volume, action, and mute. You will also get clues on what Alexa is currently doing based on the color of the light that appears on the light ring.

Front

3.3"

1.3"

amazon

| Volume | Microphone | Volume | Light | Action |
| Up | Off Button | Down | Ring | Button |

Image from Amazon.com

The action button is meant to respond to an action, depending on the context of your command – this is the button to hit when you need to turn off an alarm, wake Alexa up without having to say "Alexa," and so on. If you want to set the device up with the Wi-Fi, all you need to do is to press and hold the action button.

The mute, or "mic off" button is designed to stop Alexa from having to respond to your commands, should in any case you accidentally say its wake-up word. The light ring will turn red once this button is pressed.

Once you unbox the device, you may notice that the Echo Dot does not have any power button - it means that it is supposed to be always powered on. Don't worry about this device overheating on the corner of your living room – during idle states, the Echo Dot will sleep until it hears its "wake up" word, which is Alexa.

Image from Amazon.com

The 7-microphone array allows you to speak to Alexa from across the room, thanks to the 360-degree speaker with far-field voice recognition. The Echo Dot can receive your voice commands well within a 40-square meter room and pick up normal speaking volumes even from afar – this means that you do not have to lean closer to the Echo Dot or scream your commands out. Another plus is that it is capable of hearing you despite noise or while you are playing a background music.

How Many Echo Dots Should I Buy?

If you are planning to stock up on Echo dots and place one in every room, you might want to hold that thought if you are living in a rather small house or apartment. If you only have one floor and your home area is less than 1000 square feet, one Alexa device should be able to cover your entire space. You can also put the Echo Dot in the center of the room that you frequently use since the device can even pick up your voice through a single closed door. If you want to minimize yelling or if you want to ensure that Alexa can hear you, you can purchase the Alexa remote instead.

What's the Difference: First and the Second Gen

If you are one of those who bought the previous edition of the Echo Dot, you might notice a slight difference in the hardware between the two versions.

The first generation (which is not available anymore through Amazon) features a dedicated reset button, a LED power light, and a volume ring.

The second generation Echo Dot, on the other hand, features volume buttons, plus you will need to perform a button combination to reset the device. This generation also does not have a LED light to indicate the power.

What's Up with the Light?

Not many people will probably pay attention to the Echo Dot's light ring. However, the light ring is actually the Echo Dot's way of sending you visual cues about what is going on in the device. Here are the light ring statuses and their corresponding meanings:

- No light
  Echo dot is waiting for your command.
- Cyan lights spinning, with solid blue light
  Echo dot is booting up.
- Cyan lights pointing towards you, with solid blue light
  Alexa is still processing your request
- The ring is orange and is spinning in a clockwise direction
  Echo Dot is connecting to the Wi-Fi
- Solid plain red light
  The microphones are on mute. Press the mic button to turn them back on.

- White light
  The volume levels are being adjusted
- Violet light oscillating continuously
  There is a problem during the Wi-Fi setup. You may want to check out the Troubleshooting chapter later to address this.

Amazon Echo Dot vs. Amazon Tap vs. Amazon Echo

If you have purchased other Internet-connected speakers from Amazon, which are the Amazon Echo and the Amazon Tap, you may notice that these speakers look and function similarly. Why bother to get a different speaker when they are bound to do the same things, anyway?

What makes them differ from each other is that every speaker works better depending on the environment that they are in. Take a look at their individual specs to compare:

Category	Echo	Amazon Tap	Echo Dot (2nd Generation)
Image			
Portable	No	Yes	No
Power	Power adapter	Non-removable, rechargeable battery, charging cradle, and power adapter	Power adapter
Wi-Fi	802.11a/b/g/n Dual-band (2.4 GHz and 5 GHz)	802.11b/g/n Single-band (2.4 GHz)	802.11a/b/g/n Dual-band (2.4 GHz and 5 GHz)
Alexa Activation	Wake word Action button	Microphone (Talk) button	Wake word Action button
Speakers	Mono	Stereo	Stereo audio output for Bluetooth & audio cables
Dolby Audio	No	Yes	No

Buttons and Lights	Light Ring Action button Mute button	Front light indicators Microphone (Talk) button Power button Wi-Fi/Bluetooth button Dedicated playback buttons	Light Ring Volume buttons Action button Mute button
Bluetooth Audio Input	Yes	Yes	Yes
Bluetooth Audio Output	No	No	Yes
AUX Audio Input	No	Yes	No
AUX Audio Output	No	No	Yes
Compatible with Alexa Voice Remote	Yes	No	Yes
Accessories	Alexa Voice Remote*	Amazon Tap Sling*	3.5mm audio cable* Alexa Voice Remote* Amazon Echo Dot Case*
Media Storage	No	No	No

Image from:
https://www.amazon.com/gp/help/customer/display.html?nodeId=202009700

Judging from the above comparison, you may realize that these Alexa-powered speakers' performance may differ based on their intended use. For example, if you are thinking of bringing a speaker with you on a trip, Amazon Tap may be the best bet to have, especially since it has the stereo sound and can be connected to your other media devices via Bluetooth.

However, what makes Amazon Echo Dot a better buy is that it is lightweight and smart – it can do all the same stuff that the bigger Amazon Echo can do, and can compensate for the lack of better sound when connected to your home speakers, which you are likely to have if you are into smart devices. If you do not have speakers that connect via Bluetooth, you have the option of using the 3.5mm jack instead.

The best thing about Dot is its price point – since it is only about $40, you can set up a home environment that is full of Alexa-powered devices that can take all of your voice commands.

Registering and Deregistering Echo Dot

Registering your Echo Dot is one of the things that you need to do in order to sync it to your Amazon.com account, as well as your other online accounts that you want to automate with Alexa. Unless you connect your Echo Dot with your Alexa account, there is not much that you will be able to do with your device.

To register your account, simply go to your Alexa app (if you still do not have it, download it on your mobile device's app store), and then register your device.

If you are planning to get a second-hand Alexa device, such as the Echo Dot, you will need to see to it that the device has been deregistered. The same goes when you are planning to give or sell an Echo Dot that you have purchased using your Amazon account. The reason is simple – you simply will not be able to use an Alexa device that has been registered under another account. This means that you will not be able to assign new skills or link up other apps or devices with your Echo Dot, which pretty much makes it useless. Also, remember that only the owner can deregister an Alexa device.

If you did not check the box that says "this is a gift" when you purchased an Echo Dot, the device will automatically become registered under your Amazon account. If you are going to give that Amazon Echo as a gift, the one that you are going to receive it will not be able to register it under his or her own account. Unless you do intend to operate the device for the recipient, make sure that you go to Settings -> (your device's name) ->

9

Deregister. Once that you see the prompt that the change has been applied, you are now ready to sell or give your Echo Dot.

Chapter 2: Setting Up the Echo Dot

Before you can start using your Echo Dot and Alexa, you will need to connect your device to a Wi-Fi network and then register it using the Alexa app to your Amazon account. Follow these steps to achieve these tasks:

1. Get the Alexa app.
   The Alexa app is free and can be downloaded into the following devices:
   - iOS 7.0 and higher
   - Fire OS 2.0 and higher
   - Android 4.0 and higher

   To get the app, go to your mobile device's apps store and then search using the Alexa app. Once you have found the app, download and installed it to your device.

   Alternatively, you can also get the app by going to www.alexa.amazon.com from your computer's browser. You can use the following browsers to download the app:
   - Chrome
   - Safari
   - Microsoft Edge
   - Internet Explorer 10 and higher
   - Firefox

   Note: The following Amazon devices do not support the Alexa app:
   - 2nd Generation Kindle Fire HD 8.9
   - 1st and 2nd Generation Kindle Fire
   - 2nd Generation Kindle Fire HD 7."

2. Turn on the Echo Dot

Place the device in a central location of the room, or at least about 8 inches away from any window or wall. Plug the power adapter that comes with the package into the Echo Dot, and then insert it into a power outlet. You will see that the Echo Dot's light ring will turn blue. When the light turns orange, Alexa will greet you.

Note: Phone chargers, other USB power adapter, and USB power banks may not provide enough power to the device. To be safe, use only the dedicated Echo Dot charger that came with the package.

3. Connect Echo Dot to your Wi-Fi
Pull up the Alexa app on your phone or tablet and then follow the instructions provided to connect the Echo Dot to your internet network.

Note: Echo Dot can connect to any dual-band Wi-Fi network that has the 802.11a /b/g/n standards. The device will not be able to connect to peer-to-peer or ad-hoc networks.
Tip: If you are not able to connect Echo Dot to the internet, restart your device by unplugging and then plugging it back.

4. Start Talking to Alexa

Once your Echo Dot is connected to the internet, you are now ready to issue commands, which all begin with the "wake word" Alexa, and then proceed to speak the rest of the command normally.

Tip: Alexa is the default wake word, but you can change it using the Alexa app.

5. Connect to a speaker
While this is an optional step, you might still want to perform this step since Echo Dot's sound is only in mono.

You can either connect your Echo Dot using a 3.5mm jack or via Bluetooth.

Connecting to the Internet

There are times wherein the entire process of connecting to the internet does not happen automatically. For this reason, you will need to perform a few tasks on your Alexa app to sync it with your Echo Dot.

1. Once you are in the Alexa app, pull up the navigation panel that you will find on the left. Click on Settings.
2. Since you are adding a new device to your account, choose *Set up a new device.* If you want to update the Wi-Fi connection for your Echo Dot, select your device from the list and then click on *Update Wi-Fi.*
3. Press and hold the Action button on Echo Dot for about five seconds. Once your device's light ring turns orange, your mobile device where you are operating the Alexa app becomes connected to your Echo Dot. Afterward, you will see a list of all Wi-Fi networks that are available.
4. Choose your Wi-Fi network and enter your password when prompted. If you don't see your connection's SSID (the Wi-Fi's name) on the list, select *Rescan.* If you still do not see it, you may need to manually connect your device to your connection. To do that, scroll down the list of available networks until you see *Add a Network.* Select this option and enter the SSID and the password of your Wi-Fi network.
5. Choose *Connect.* Once Echo Dot is connected to the internet, you will see a confirmation message. You are now ready to talk to Alexa!

There are situations wherein you will want to bring your Echo Dot with you outdoors – since this device is reliant on the Internet, you will need to connect it to your mobile phone's hotspot.

To connect Echo Dot to the internet via a hotspot, see to it that you have the following:

- A cellular phone plan that will enable you to turn on hotspot for a considerable amount of time
- An Echo Dot that has a software version of 3389 or higher
- An Alexa app on your mobile device.

Special Note: When you used your Echo Note for the first time, you may have noticed that you do not have the option to connect to a Wi-Fi hotspot. To enable this feature, you will need to see to it that you have downloaded and installed the latest software update to get this feature on your device.

Follow these steps to connect your Echo Dot to a hotspot:

1. Pull up your mobile device's menu and look for the Wi-Fi hotspot option. Copy the SSID or network name and the password of your device's hotspot network
2. Pull up the Alexa app, and then choose Settings on the app's left navigation panel
3. Select your device and then choose Update Wi-Fi
4. Press and hold the Action button on your Echo Dot. When the light on the light ring becomes orange, your Echo Dot will be connected to your mobile phone. Take a look at your Alexa app, and you will see the Wi-Fi networks that are currently available.
5. Scroll down and select the option Use this device as a Wi-Fi hotspot
6. Choose Start.

7. Enter your device's SSID and the corresponding password. Afterward, select the option Connect.
8. Turn on the Wi-Fi hotspot on your mobile device. Alexa will confirm that you have made a successful connection.

Connecting Echo Dot to Speakers Using Aux Out

If you have an audio cable with a 35mm jack, then you can connect your Echo Dot to an external speaker. All you need to do is to follow these steps:

1. Place your Echo Dot at least three feet away from your speakers. When your Echo Dot is too close to your speakers, there is a risk that Alexa may not hear her wake up call as well as other commands.
2. Turn your speaker on.
3. Plug in one end of the cable on your Echo Dot and the other end on your speakers. Doing so will stream all audio to your speakers.

Connecting Echo Dot to a Bluetooth Speaker

Since the Echo Dot does not have a bigger can of speaker like its bigger brother Amazon Echo, it will make sense to connect it to a better speaker to hear Alexa's responses better or to simply enjoy better audio.

Before you start this setup, see to it that you do the following steps first:

- Make sure that your Bluetooth speakers and the Echo Dot are at least three feet away from each other. The reason is that when the Echo Dot is too close to the speakers, Alexa may have a hard time hearing the wake word and commands.
- See to it that your speakers can connect to other devices with Bluetooth, such as your mobile phone.
- Check the Bluetooth speaker by turning it on and adjusting the volume.

- Disconnect other Bluetooth devices that may still be connected to your Echo Dot. The device can only accommodate one Bluetooth connection at a time.

Follow these steps to connect your Echo Dot to a Bluetooth speaker

1. On your Bluetooth speaker, switch the pairing mode on.
2. Pull up the Alexa app on your phone or tablet and then choose Settings
3. Choose Echo Dot from the list and then navigate to Bluetooth -> Pair a New Device. This will allow your Echo Dot to enter the pairing mode and search for Bluetooth devices. Once your Echo Dot detects your Bluetooth speaker, the speaker will appear on a list that displays available devices for pairing on the Alexa app.
4. Select your speaker from the list. Once done, Echo Dot will connect with the Bluetooth speaker. Alexa will tell you if the pairing was successful. Select Continue on the Alexa app.

Tips:

- If you want to connect with the last Bluetooth device that your Echo Dot paired with, say "Connect."
- If you want to pair with a different speaker, say "Disconnect."
- If you want to manage all Bluetooth devices that you are using with Alexa, pull up the Alexa app and then navigate to Settings -> Device Name -> Bluetooth

Pairing Echo Dot with Your Mobile Device

One of the wonders that you can do with Echo Dot is that you can use it to control almost every app that you have with your mobile device. If you want to check someone's profile on Facebook by using your voice, you will need to first pair your Echo Dot with your smartphone by following these steps:

- Disconnect any other Bluetooth device that is paired with Echo Dot.
- Turn on your mobile device and then set it on pairing mode. If possible, place it closer to your Echo Dot
- Pull up the Alexa app on your mobile phone.
- Go to the left navigation panel, and then choose Settings
- Select your Echo Dot, and then go to Bluetooth -> Pair a New Device. This will make your Echo Dot enter pairing mode.
- Open the Bluetooth settings on your mobile device and the choose Echo Dot. Alexa will tell you if you have successfully connected Echo Dot to your mobile device.

Chapter 3: Dealing with Alexa

Alexa is the voice service that allows you to communicate with your Echo Dot, as well as Amazon Echo and Tap. You can think of Alexa as similar to Siri – with the combination of the "wake" word Alexa and your command; you can use this voice service to answer questions, search the web, schedule an alarm and so much more. Most Echo Dot reviews also tell that Alexa is far more responsive and natural-sounding than other phone voice assistants out there, so if you are using other voice-activated apps to manage your online accounts or to serve as a remote control, you may find yourself using your Echo Dot more frequently than your phone.

What makes Alexa special is that it is cloud-based, which means that it constantly received new information and is therefore bound to get smarter. The more customers use Alexa-driven products, such as the Echo Dot, the more she gets to work with different sets of vocabulary, speech patterns, and personal preferences. Because Alexa also works well with most smart home appliances, this voice assistant will soon be your smart home's de facto voice assistant.

How Does Alexa Work?

Alexa (aptly named after the Library of Alexandria) works using Amazon's system for language processing, which is deemed to among the easiest systems to interact with. This means that you will rarely find yourself having to issue a command or ask your question twice. You can count on Alexa to always be listening and quick to respond.

Your voice is streamed to the cloud whenever you issue a voice command, which also shows on your Alexa app. When you ask a question or ask Alexa to do something, Echo Dot will light up and then send the query or command to Amazon's cloud database to be analyzed. If you have integrated most of your apps and smart devices to your Alexa device, you will be able to maximize your Echo Dot's use.

The Amazon Alexa App

The Alexa App, which you can download for free from US app stores, allows you to integrate Amazon devices such as the Echo Dot to your other apps and smart devices. The app also comes with quite some skills, such as finding books and music, controlling the music tracks that you are playing on your Echo Dot, setting to-do lists and alarms, and so much more. It also allows you to discover and activate other voice-driven skills.

Navigating the Alexa App

The Home Screen on your Alexa app tells you what activity Alexa is currently doing. When you navigate through the cards on this screen, you will be able to see your recent activities and different other features, such as to provide feedback or to remove a particular card.

When you click on the app's menu icon, you will be able to pull up the left navigation panel. This panel will allow you to see other settings and available features on your Echo Dot.

Tip: When you pull up the Alexa app on a computer, you will see that the navigation panel is already up.

Here are the items that are available on the Menu:

Menu Option	Description
Home	View your recent interactions with Alexa.
Now Playing	View and control tracks playing on your device, show upcoming tracks in the queue, and view your history.
Music & Books	Find songs, stations, shows, Kindle books, and audiobooks to listen to on your Alexa device.
Shopping & To-do Lists	View and manage your shopping and to-do lists.
Timers & Alarms	View and manage timers and alarms.
Skills	Find and turn on skills that add various voice-driven capabilities to your Alexa device.
Smart Home	Manage smart home devices linked to Alexa.
Things to Try	View a list of example phrases you can say to Alexa.
Settings	Set up your new Alexa device, train Alexa to understand your speech patterns, and adjust various Alexa device settings.
Help & Feedback	Find detailed help for your Alexa device and submit feedback about your experience with Alexa.
Not [NAME]? Sign out	Sign out of the Alexa app on your device.

Image from: Amazon.com

Amazon Alexa's Device Settings

The Alexa app allows you to make software and hardware changes to your device, such as naming your device or changing the registration information for your Echo Dot. You will also be able to find other important information about your device under these settings, which will allow you to make better use of your Echo Dot, especially if you want to use it with other smart appliances in your home.

- Update Wi-Fi or Connect to Wi-Fi

This setting allows you to connect to a wireless connection or to make changes in your existing connection

- Bluetooth
This option allows you to remove and add a Bluetooth connection on your Echo Dot. Keep in mind that you can only have one Bluetooth connection to your device, so make sure that you check if your Echo Dot is connected to another device before attempting to pair it with another gadget.

- Sounds
This option allows you to browse and select sounds that you want your Echo Dot to make whenever you interact with it.

- Device Location
This option allows you to let your Echo Dot know your location. When you add your address, you will be able to ask Alexa about movie schedules and other places near you. You will also be able to get real-time information about the traffic, weather, or ask for the current time.

To change your location on your device, pull up the left navigation panel, and then choose Settings. Select your device, and then tap on Device Location -> Edit. Fill up the fields by entering your street address, city, state, and your ZIP code. Hit Save after entering the information.

Tip: If you have multiple Echo Dots in your household, make sure that you update the location of each device.

- Device Name
This option allows you to provide a name for each Alexa device that you own. Doing so will be extra handy especially if you have different Alexa devices in your household.

To change the name for your device, enter the desired name in the Name Field, and then tap Save.

- Wake Word
This option will allow you to change the wake word to which your Echo Dot responds to. This will be helpful if you have more than one Echo Dot in a room – by changing the wake word, you can prevent multiple Echo Dots from responding to a command.

  To change the wake word, go to Settings, and then select your device. Tap Wake Word, and then choose your desired wake word from the menu. When you are done, tap Save. Your Echo Dot's light ring will turn orange for a moment, and then the device will begin to respond to the new wake word.

  Note: You are allowed to use the wake words Amazon, Echo, and Alexa.

- Metric Measurements
This option allows you to use the metric system to measure distance and temperature.

- MAC Address
This will show you your Echo Dot's MAC address.

- Device Software Version
This will show you the software version that is installed on your device. Your Echo Dot should automatically update its software using your internet connection. If you want to make sure that your device has the latest version installed, go to Settings -> Device Software Version, and then check if it has the software version 564196920. If it shows that it has an older version, unplug the power adapter and then plug it back in. Once it is connected to your internet

connection, it should automatically download the latest version. Your Echo Dot's light ring will turn blue and will then begin to install the latest software.

Tip: It may take up to 15 minutes for your device to download a new software version, depending on the speed of your internet connection. To prevent any interruption, refrain from saying the wake word or interacting with your Echo Dot while it is doing a software update.

- Serial Number
  This will display your Echo Dot's serial number

- Device is Registered To
  This option will show you your Echo Dot's registration status. If you are planning on giving your Echo Dot to someone else, make sure that you select Deregister for the new user to be able to register the device under his or her account.

- Pair/Forget Remote
  This option will allow you to connect or disconnect an Amazon Echo remote (this item is sold separately) to your Echo Dot. Once you have already paired your remote (to do this, go to Settings -> Pair Remote), hold down the button labeled Play/Pause on your remote control for about five seconds. Once you release the button, your Echo Dot will search for the remote and then attempt to connect with it. Once your Echo Dot has found the Alexa remote, your voice assistant will prompt you with a pairing success message.

The Echo Dot's Account Settings essentially lets you perform activities using your device and activate certain skills that you want it to perform. Making some changes to the Account settings will also teach your Echo Dot to respond better to your commands and sync to your other apps, accounts, and compatible smart devices.

- Voice Training
  This option allows Alexa to improve its ability to understand your speech patterns and respond to your commands better. When you choose to do a Voice Training, you will be prompted to read 25 phrases from your Alexa app to your Echo Dot. Your Echo Dot will then process each of the phrases you say.

  To do voice training, pull up Settings -> Voice Training -> Start Session. Proceed to read a phrase from the app, and then tap Next to go to the next phrase. If you want to redo a phrase, tap Pause, and then select Repeat Phrase. Once you have completed your training session, tap on Complete.

  Tip: You can choose not to read all 25 phrases in a training session. If you want to quit Voice Training, simple choose the Pause option and then tap End Session.

  Note: For best results, make sure that you position yourself in a spot where you would normally speak to your Echo Dot. Train it to respond to your normal speaking voice. If you purchased an Alexa remote, do not use it while you are doing voice training to teach your device to respond to you without the need to shout.

- Calendar

This option allows you to link your Google Calendar to Alexa. Once these two applications are synced, you can ask Alexa to tell you upcoming events.

- Traffic
  This option allows you to ask Alexa about your travel time to a particular destination

- Music and Media
  This option lets you connect Alexa to media or music service.

- Flash Briefing
  This option lets Alexa tell you about weather updates, news headlines, or choose shows that you want to tune into.

- Voice Purchasing
  This option allows Alexa to order products from Amazon, such as physical goods, e-books, or digital music. You will also be able to use a confirmation code that Alexa will prompt you to say before you place your orders.

- Smart Home
  This option allows you to pair smart devices with your Echo Dot and allow Alexa to control them using your voice.

- Household Profiles
  This option allows you to join or create your own Amazon Household in order to hear content from different Amazon accounts.

- About the Amazon Alexa App
  Selecting this will display the latest Alexa app version.

Now that you have an idea of the different things that you can do with Alexa, it's time to discover how you can make full use of your Echo Dot.

Chapter 4: Listen to Music and Media with Your Echo Dot

Since your Echo Dot is essentially a voice assistant, one of the best things that you can do with it is to stream audiobooks, music, podcasts, and other types of audio content from your other devices that you can connect via Bluetooth. You can also use your Echo Dot to listen to your own collections from Google Play Music, iTunes, or your media library on your desktop. This chapter will teach you how you can use your Echo Dot to stream audio content from different resources.

Commands for Playback

If you are playing media through Bluetooth, or in your Amazon Music Library, you can control your playback by using your selected wake word plus the following words:

- Stop
- Play
- Resume
- Set volume to level [desired number]
- Restart
- Volume up /volume down
- Loop
- Repeat
- Skip
- Play some music
- Plat the album [album title]
- Play some [genre] music
- Play songs by [artist name]
- Play some [name of genre] from Prime

- Listen to my [playlist name] playlist
- Play [name of station]
- Play [name of holiday] music

If you want to skip tracks in the playlist or album you are currently playing, say the wake word plus the following words:

- Next
- Previous
- Shuffle/stop shuffle
- Shuffle my [playlist title] playlist

If you want to hear more information about the song that your Echo Dot is playing, say the wake word, plus the following:

- Who is this artist?
- What song is this?
- What is this?
- When did this song (or album) come out?
- Who's the lead singer from [name of group]?

There are also some commands that are specific to a music service where you are playing a song from:

1. If you want to play tracks from Prime Music, say the wake word plus the following:
   - Play some Prime Music
   - Play [name of station] from Prime
   - Play [song, artist, or album] from Prime Music
   - Play a Prime playlist
   - Play [name of playlist] from Prime
   - Play [name of genre] from Prime Music
2. If you want to play tracks from Spotify Premium, say the wake word plus the following:
   - Play [song] from Spotify
   - Play songs by [name of artist] from Spotify
   - Play [name of playlist] from Spotify

- Play Spotify
- Play [song title] by [artist name] from Spotify
- Connect to Spotify / Spotify Connect (this command will enable Spotify Connect)

3. To play a custom station on Amazon, iHeartRadio and Pandora, say the wake word plus the following:
   - Play my [genre/artist name] station on [iHeartRadio/Prime Music/Pandora]

4. To play a radio station in iHeartRadio and TuneIn, say the wake word plus the following:
   - Play the station [name of station]
   - Play [frequency of selected station]
   - Play [station's call sign]

5. To play a program or podcast in iHeartRadio or TuneIn, say the wake word plus the following:
   - Play the program [program title]
   - Play the podcast [podcast title]

6. To like or dislike a song in Amazon Music, iHeartRadio, and Pandora. Say the wake word plus the following:
   - I [like/don't like] this song
   - Thumbs up/down

7. To remove a frequently-played track from the queue in Amazon, iHeartRadio, and Pandora stations, say the wake word plus the following:
   - I'm tired of this song

8. To create a station in Pandora and iHeartRadio, say the wake word plus the following:
   - Create [Pandora/iHeartRadio] station based on [name of artist]
   - Make a station for [name of artist]

Music and Media Services You Can Use

Alexa is able to stream content to your Echo Dot from these services:

- TuneIn
- Spotify
- Amazon Music
- Pandora
- Prime Music
- Audible

In order to stream music from these services to your Echo Dot, all you need is to ask Alexa to play music directly from these services. You automatically have access to any music track that you have in your Amazon Music library. You will also be able to automatically access audiobooks that you have in your Audible library using your Echo Dot. If you have an Amazon Prime membership, you can stream entire stations, playlists, and about two million free songs.

Tip:

To set your default music service preferences, pull up the Alexa app, and then go to Settings -> Music & Media -> Choose default music services. This option will allow you to choose your default music library (Spotify or Amazon Music) and your default station service (Pandora, Amazon Music, or iHeartRadio). Tap Done after choosing the default services that you want.

Linking Third Party Music Accounts

However, you should take note that you may need to link your accounts or subscriptions (i.e. Pandora, Spotify Premium, and iHeartRadio) your Amazon account via the Alexa app in order to command your Echo Dot to play audio tracks from these services. To link these third-party services to Alexa, follow these steps:

1. Pull up your Alexa app.

2. Go to the left navigation panel and tap Settings
3. Choose the Music & Media option and then select the streaming service you want to sync with Alexa.
4. Tap the option Link account to Alexa. You will see a sign-in page to the third-party service that you have selected.
5. Enter the email address and the password for the third-party service. (Note: This is not the same email and password that you are using for your Amazon account.)

If you want to unlink the streaming service from Alexa, simply select the Unlink account from Alexa option in Settings.

Note: You can use the Spotify app on your mobile device as a remote control if you are using Spotify Connect.

How to Upload Your Music

One of the most common questions that Echo Dot owners have is this: Is it possible to make my Echo Dot play the music that I have on my Google Play account or iTunes? Right now, it is not possible for Alexa to link to these media services. However, your voice assistant can instead pair your Echo Dot to another device that is linked to these services directly so you can enjoy some playback functions. For example, if you want to connect your laptop to Echo Dot, all you need to do is to enable Bluetooth for both devices and say this command:

Alexa (or your selected wake word), pair.

Alexa will then find your device and then tell you to go to the device that you want to pair with your Echo Dot and select the option Echo from the pairing screen. Once both devices are linked together, you can pull up the music app on your laptop or smartphone and play some music.

However, if you want to get the same functionality that you enjoy when you are playing music from your Amazon Music library, you can upload your music collection from your Google Play account or iTunes library on your Amazon Music library instead. To do that, you can use the Amazon Music (available on PC and Mac) to upload your music tracks straight from your computer.

After uploading, you can simply command Alexa to control audio playback and the tracks that you are playing.

You are allowed to upload a maximum of 250 tracks to My Music without having to pay for anything. However, if you want to upload more songs, you can pay for an Amazon Music subscription, which will allow you to upload up to 250,000 songs. To do that, sign in to your Amazon account and head over to the option Your Amazon Music Settings. Once you are there, choose the option Upgrade music library storage. Enter your preferred payment method then hit Continue. Once you have an upgraded storage subscription, you can immediately use the additional storage that you have purchased.

Tip: If you made a purchase from the Digital Music Store or if you purchased Amazon vinyl and CD records that are eligible for an AutoRip, those tracks do not count towards your 250 track limit.

Preventing Drops in Bluetooth Playback

Take note that the commands that were listed above are both used for Bluetooth playback and in playing music that you have in your Prime or Amazon Music library. This means that it is possible that when you use one of these commands during a playback using Bluetooth, Alexa may pause the playback that you are doing from your smartphone or computer, disconnect, and then play the music that you want from your Amazon Library instead.

Here is an example: if you are currently playing the album Sea Saw by Lisa Hannigan from iTunes on your Echo Dot using a Bluetooth connection, then decide that you want to play Crossroads by Bone Thugs and Harmony from the same playlist, you may find yourself saying this command:

[wake word], play Crossroads Bone Thugs and Harmony

What will happen next in this scenario is this: Alexa will pause the Lisa Hannigan song that was playing, then disconnect the Bluetooth device. Afterward, it will attempt to find the song that you requested in the Amazon Music library. If you want to resume playing the song that you want via Bluetooth instead, what you can do is to reconnect your computer or smartphone by saying:

[wake word], connect

Your Echo Dot will re-connect with the last Bluetooth device it was paired with, and then you can select the track that you want and hit Play.

How to Listen to Audio Books

Alexa offers support for the software Whispersync for Voice, which allows you to keep the current position when an audiobook is playing. This means that you can listen to an audiobook on your Echo Dot and then continue listening from where you left off from any Amazon app or device. If you want to listen to an audiobook that you do not have, Alexa will play a short passage from the start of an Audible book. You can listen to audiobooks if you have an existing Kindle Unlimited or Audible subscription.

Note:

Alexa does not offer support for these Audible features/content:

- Notes
- Badges and stats
- Bookmarks
- Speed control for the narration
- Magazine and newspaper audio content

Here are the commands that you can use to control playback for audiobooks:

1. To listen to an audiobook, say the wake word plus the following:
   - Read [audiobook title]
   - Play [audiobook title] from Audible
   - Play the book [audiobook title]
   - Play the audiobook [audiobook title]
2. To pause reading, say the wake word plus the following:
   - Pause
3. To have Alexa continue reading your most recent book, say the wake word plus the following:
   - Resume my book
4. To go forward or backward in Alexa's reading by 30 seconds, say the wake word plus the following:
   - Go forward
   - Go back
5. To go to the previous or next chapter, say the wake word plus the following:
   - Previous chapter
   - Next chapter
6. To jump to a particular chapter, say the wake word plus the following:
   - Go to chapter [chapter number]
7. To restart a chapter, say the wake word plus the following:
   - Restart
8. To cancel or start a timer for sleep, say the wake word plus the following:
   - Cancel sleep timer
   - Set a sleep timer for [set number] hours/minutes
   - Stop reading the book in [set number] hours/minutes

Reading Kindle Books with Echo Dot

Alexa is capable of reading Kindle books that you have in your library using the same technology that it uses for reading calendar events, news, and Wikipedia articles. You can ask Alexa to read the following Kindle content:

- Items that you have bought from Kindle Store
- Content lent by Kindle Owner's Lending Library
- Content that other Kindle owners shared in your Family Library
- Content lent by Kindle Unlimited.

If you want to look for books that Alexa can read using your Alexa app, pull up the navigation panel and then tap Music and Books. Afterward, navigate to Kindle Books -> Books Alexa Can Read.

Note: Alexa will not be able to do the following:

- Perform immersion reading
- Perform speed control for narration
- Read graphic novels and comics

Just like in reading e-books, Alexa can pick up from the page where you left off in a book that you were reading a book that you have in a different Amazon reading device or e-book reading app.

Pro Tip: If you want to jump to a different chapter in a Kindle book, tap Now Playing in the Alexa app as your Echo Dot reads you a Kindle e-book, then choose Queue. You will be able to see a list displaying all the chapters that you can select from.

Alexa Commands for Kindle Reading
Here are the commands that you can say to Alexa to control your Kindle book reading:

1. To listen to a Kindle book, say the wake word plus the following:
   - Read [title of Kindle book]

- Read my Kindle book
- Play the Kindle book [title of Kindle book]
- Read my book [title of Kindle book]
2. To pause the Kindle book playing on your Echo Dot, say the wake word plus the following:
   - Stop
   - Pause
3. To continue playing your Kindle book, say the wake word plus the following:
   - Resume
   - Play
4. To go to the previous or next paragraph in the Kindle book that is currently playing, say the wake word plus the following:
   - Go back
   - Skip back
   - Skip ahead
   - Go forward
   - Previous
   - Next

Chapter 5: Listening to Traffic, Weather, and News with Your Echo Dot

Among the features that Echo Dot owners love about their device is the Flash Briefing, which allows them to hear news-in-brief content from different websites. The National Public Radio, or NPR in short, is the default source for briefings that you will hear from Alexa. However, you can change Alexa's settings in the mobile app to get your briefings instead from other sources such as TMZ, The Economist, AccuWeather, or ESPN.

How to Change Flash Briefing Settings

If you want to change the content that you hear whenever you say "Alexa, Flash Briefing," follow these steps (you might want to pull up your Alexa app on your mobile phone as you read along):

1. Tap Settings on the left navigation panel.
2. Select Flash Briefing to pull up the Flash Briefing menu.
3. From the Flash Briefing menu, select or deselect sources by tapping toggle control that you will see on the right.
4. Scroll down to refine the topics that you will hear from your Echo Dot (topics will also have on/off toggle control).

Alexa Commands for Flash Briefing
Here are the commands that you can use to control your Flash Briefing playback:

1. To listen to your Flash Briefing, say the wake word plus the following:
   - What's new?
   - What's my Flash Briefing?

2. To navigate through your Flash Briefing, say the wake word plus the following:
   - Cancel
   - Next
   - Previous

How to Get Sports Updates

You can listen to the latest sports news and get upcoming game schedules and hear the latest scores using Alexa's Sports Update. All you need to do is say "Alexa (or your selected wake word), give me my Sports Update." To get sports news about your favorite sports team, you will need to add the name of the team that you want to follow in the Alexa app. Here is how you can do it:

1. Pull up your Alexa app and navigate to Menu -> Settings -> Sports Update
2. Type the name of your favorite team in the search field. You will see suggested teams in the field as you type.
3. Choose the name of the sports team you want to add.

Tip: If you want to remove a sports team, tap the X that will appear next to the name of a sports team in the app. You can add up to 15 sports teams for your Sports Update.

Alexa's Supported Leagues

To date, you can hear sports updates about the following sports leagues:

- NBA National Basketball Association
- BL Bundesliga
- MLS Major League Soccer
- EPL English Premier League
- WNBA Women's National Basketball Association
- NFL National Football League
- NCAA National Collegiate Athletic Association men's basketball
- WNBA Women's National Basketball Association
- NHL National Hockey League

Checking the Weather with your Echo Dot

You can ask Alexa about the weather in your vicinity, as well as forecasts in any city US or any city around the globe. When you ask about a weather forecast, the Alexa app will show you a card that will show you a forecast for seven days for the location that you requested from AccuWeather.

Tip: Don't forget to add your address in your Alexa app to get local forecasts.

Alexa Commands for Weather Forecast
Here are the commands that you can use to control your weather forecast playback:

1. To ask about the weather in your current location, say the wake word plus the following:
    • What's the weather
2. To hear about the weather on a particular day, say the wake word plus the following:
    • What's the weather for [specific day]?
    • What's the weather for this weekend?
    • What's the weather for this week?
3. To hear the weather forecast in a different city, say the wake word plus the following:
    • What's the weather in [city and country/city and state]?
4. To hear about any inclement weather conditions, say the wake word plus the following:
    • Will it be windy tomorrow?
    • Will it [snow/rain] tomorrow?

How to Check Traffic Using Echo Dot

If you are headed to a particular destination and you want to check the traffic conditions for your commute, Alexa will be able to provide the best route that you can take and the estimated time of your arrival. To setup Alexa to provide commute information, follow these steps:

1. Pull up Settings-> Traffic from the left navigation panel in the Alexa app.
2. Go to the From and sections and tap Change address.
3. Input your location and the destination address in the fields and then tap Save changes.

Pro Tip: If you need to add a stop to your commute route, tap Add stop. Take note that you can only add a single stop to your commute route.

Alexa Commands to Check Traffic
Here are the commands that you can use to check traffic:

1. To hear traffic update, say the wake word plus the following:
   - What's my commute?
   - How is traffic?
   - What's traffic like right now?

How to Locate Nearby Places

You can use your Echo Dot to search for businesses, restaurants, or shops that are near your location. Alexa makes use of your location to search Yelp for nearby locations.

Alexa Commands for Checking Nearby Locations
Here are the commands that you can use to find nearby establishments and other information using your Echo Dot:

1. To look for different types of restaurants or establishments nearby, say the wake word plus the following:
   - What [restaurants/businesses] are close by?
   - What [restaurants/businesses] are nearby?
2. To look for restaurants or establishments that have good ratings on Yelp, say the wake word plus the following:
   - What are some top-rated [restaurants/businesses]?
3. To find out the address of a restaurant or business that is near your location, say the wake word plus the following:
   - Find the address for a nearby [restaurant/business].
4. To find the phone number of a restaurant or business that is near your location, say the wake word plus the following:
   - Find the phone number for a nearby [restaurant/business].
5. To find out the operation hours of a nearby restaurant or business, say the wake word plus the following:
   - Find the hours for a nearby [restaurant/business].
6. To ask about further information about a business that Alexa has provided you, such as operating hours, phone numbers, or street address, say the wake word plus the following:
   - How far is it?
   - What is the phone number?
   - Are they open?

Use Your Echo Dot to Know Movie Schedules

You can ask Alexa to provide information about movies that are screening in nearby theaters or in another city. Alexa will use the address that you have in your device and IMDb to provide you information about theaters and the movies that are currently playing. You can also ask additional information about the movies that you are inquiring about, such as critic ratings, when you make a request.

Alexa Commands to Ask About Shows

Here are the commands that you can use with your Echo Dot to find information about screenings and theaters nearby:

1. To know which movies are being shown in theaters, say the wake word plus the following:
   - What movies are playing?
2. To know which movies are playing in a particular city, say the wake word plus the following:
   - What movies are playing in [name of city]?
3. To know which movie genres are showing in cinemas, say the wake word plus the following:
   - What [name of genre] are playing?
4. To know showtimes for a movie you want to watch, say the wake word plus the following:
   - When is [title of movie] playing [this weekend / tomorrow / today]?
   - What time is [title of movie] playing
   - What movies are playing between [specific times]
5. To learn more about a movie that you want to watch, say the wake word plus the following:
   - Tell me about the movie [title of movie]
6. To find out screening times at a particular movie theater, say the wake word plus the following:
   - When is [title of movie] playing [this weekend / tomorrow / today] at [name of movie house]?
   - What movies are playing at [name of movie house]?
   - What are the show times for [title of movie] playing at [name of movie house] in [name of city]?
   - What time is [title of movie] playing at [name of movie house]?

Pro Tip: If you use Alexa's Uber skill, you may get a ride to a theater in some locations. Once you are able to get information about the theater where you want to watch a movie, say "Get me a ride to [theater's location]".

Introducing the Flash Briefing Skill API

As you already know, the Flash Briefing Skill allows you to hear content. However, you can also have other Echo Dot owners hear content that you can create. This will be extra helpful if you have a business that would like to roll out special news for customers on a daily basis. If you have the Flash Briefing Skill API, you can provide a text content in the Alexa App that your customers can read or an audible content that can be played via the Echo Dot.

Who Can Make Flash Briefing Skills?

If you have an original content (or the right to distribute) that can be updated on a regular basis and you have the ability to design RSS, HTTPS, or JSON content, then you can create customized Flash Briefing content for other users. To create a customized Flash Briefing skill, you will need the following:

- A developer account for Amazon (you can get this for free)
- Any Alexa device that you can use for testing your custom skill
- Ability to modify RSS or JSON content

Chapter 6: Ask Alexa Something

Just like Siri, you can ask Alexa about almost anything that exists on the Internet, such as music, sports, people, and geography. You can also ask Alexa to convert measurement units, do some math, define unknown words, or spell a word out for you.

Note: While Alexa may be getting information using the internet, there are some situations wherein Alexa may not be able to provide you the answer to your question. To improve Alexa's services, you may use the Alexa app to let the developers know that Alexa does not have an answer to a specific question.

Alexa will also be able to make mathematical calculations, but right now, it can only perform one operation at a time. For example, you can say "[wake word], what is one plus one?" but you cannot ask Alexa to perform a more complex computation such as 1+3+5 or (7-4) x 10. The mathematical operations that Alexa can perform include the following: power, addition, division, subtraction, multiplication, square root, and factorial.

Alexa Commands for Asking Information

1. To ask about a particular person, say the wake word plus the following:
   - Who is [name of person]?
   - Who is [government position, etc]?
2. To ask about a particular date in history, say the wake word plus the following:
   - When is [name of holiday]?
   - When did [event in history] happen?
3. To ask details about movies or TV shows, say the wake word plus the following:
   - Who starred in [name of movie or TV show]?
   - What year did the TV show/movie [show/movie title] come out?

- Who stars in [name of show/movie]?
- What is [name of actor or actress] latest movie/TV show?
- What is the MDb rating for [name of show/movie]

4. To ask about a song or a music artist, say the wake word plus the following:
   - Who sings the song [song title]?
   - What year did [name of artist] release [album or song title]?
   - Who is in the band [name of band]?

5. To ask how a word is spelled, say the wake word plus the following:
   - How do you spell [word you want Alexa to spell out]?
     Pro tip: if you want to use Alexa to give you some ideas for a Spelling Bee contest, you can say the wake word plus the following: Give me a Spelling Bee word.

6. To ask what a word means, say the wake word plus the following:
   - What is the definition or [word you want Alexa to define]

7. To convert measurements from one system to another, say the wake word plus the following:
   - How many [type of unit] are in a [type of unit]?
   - How many [type of unit] are in [number] [type of unit]?

8. To ask questions about geography, say the wake word plus the following:
   - What is the capital of [name of place]?
   - Which [countries/states] border [name of place]?
   - What is the elevation of [name of place]?
   - What is the latitude and longitude of [name of place]?

9. To ask about information about a particular type of food, say the wake word plus the following:

- How many calories are in [type of food]?

10. To ask about the time in a particular location, say the wake word plus the following:
    - What time is it in [name of city]?

11. To hear a Wikipedia article about a particular topic, say the wake word plus the following:
    - Wikipedia [category or topic that you want to hear about].

Note: You will not hear the entire article, but you can say "Hear more," "More," or "Tell me more" if you want Alexa to read more of the article.

Chapter 7: How to Be Productive with Your Echo Dot

Apart from using your Echo Dot to find information online or play music, you can make your household productive by using it to schedule alarms or timers, create lists, tell you your things-to-do, or even work with third-party productivity apps.

Managing Your Alarms or Timers

You can use Alexa to create multiple alarms or timers – if you have other Echo Dots in your household, take note that the alarms and timers that you have set on one device does not automatically become set for the other Echo Dots that you have. All timers and alarms for every device is independent even if they are registered on a single Amazon account. Also take note that even if you mute your device or disconnect it from the internet, all your alarms and timers will still go off.

You can create up to 100 alarms and timers using your Alexa app. If you want to set a single alarm or timer, you can use your app to set it up to 24 hours ahead.

Setting Timers with Alexa

If you want to set a timer in your Echo Dot, all you need to do is say the wake word, and then "Set timer for [desired duration]. Once you have set a timer, you can ask Alexa to tell you how much time you have left on the timer. You can also use your voice to command Alexa to stop, cancel, pause, or resume the countdown you have set.

If you want to manage a timer that you have set, pull up the Alexa app on your mobile phone and then follow these steps:

1. From the Menu, go to Timers and Alarms. Select the device that has the timer you want to manage.
2. Tap Timers to see the timer status.
3. Find the timer that you want to manage and then tap Edit. You can choose to Cancel or Pause the selected timer.

Setting Up Your Alarms

You can tell Alexa to set an alarm by simply saying "Set an alarm for [time]." If you want to set a repeating alarm, you can say "Set a repeating alarm for [day] at [time]."

If you want your Echo Dot to snooze while it is sounding, simply say "snooze". Your alarm will give you nine minutes before it resumes sounding off again.

If you want to manage an alarm that you have already set, pull up your Alexa app and then follow these steps:

1. From the Menu, tap Timers & Alarms
2. Select the device that contains your alarm.
3. Tap Alarms and choose the alarm that you want to edit.
4. Select the option that you desire under the Repeats tab:
   a. Everyday
   b. Weekend
   c. Never repeat
   d. Every [day of the week]
   e. Delete alarm
   f. Weekends
5. Once you are done editing, tap Save Settings.

Change Sound and Volume for Alarms and Timers

If you want to change the alarm sound and volume for an Echo Dot, pull up the Alexa app and follow these steps:

1. From the Menu, choose Timers & Alarms
2. Choose the device that contains the alarm or timer that you want to edit
3. Select from these options

a. For timers: Tap Manage timer volume
b. For alarms: tap Manage alarm volume and default sound, or choose an alarm and then tap Alarm sound.

To change the sound volume for your timers and alarms, navigate to Settings -> [name of your device] -> Sounds -> Alarm and Timer Volume.

Creating and Managing Lists

If you have some important tasks or things to buy that you need to be reminded of, your Echo Dot can be a great assistant that you can ask about items in your to-do or shopping lists.

If you have created a list, you will be able to see up to 100 items on each list that you have created. You can create list items that are up to 256 characters. If you want to print your list, all you need to do is to view them on your computer's browser and then print your list from there.

You can view and access your to-do and shopping lists in your Alexa app, the Amazon app, the Amazon Website, and on the Alexa Shopping list. You can also make use of third-party productivity services that you can link with Alexa so you can access and manage lists in those other apps.

Alexa Commands to Manage Lists
To manage lists using voice commands, follow these instructions:

1. If you want to add items to lists that you have, say the wake word plus the following:
   • Add [name of item] to my Shopping List.
   • Put [name of task] on my To-do List.
2. If you want to hear what you have on your lists, say the wake word plus the following:
   • What's on my To-do List?

- What's on my Shopping List?

To manage lists that you have on Amazon or in the Alexa app, follow these instructions:

1. To print a list, pull up the Alexa app in your computer and choose Shopping & To-do Lists. Select a list that you want to print and then hit Print. If you have your list on your Amazon account, pull up the list that you want to print out and then use the printing options available on your web browser.

2. To pull up a list that you have already created, pull up the Alexa app and tap the Shopping & To-do Lists option in the left navigation panel.

   Pro Tip: If you do not have an internet connection and you need to read your To-do or shopping lists, you will still be able to view them on your smartphone or tablet using the Alexa app.

   If you want to view your lists on the Amazon app or website, navigate to Your Lists -> Alexa Shopping List.

3. To add items to your existing lists, pull up the Alexa app and find the list that you want to edit. Enter the task or item that you want to include in your list and then tap the + icon.

   To edit your lists on the Amazon app or website, simply type the item that you want to include on the available text field, and then hit the Add (+) or Add to Shopping List button.

4. If you want to edit an item that is included in your list, pull up the Alexa app and select the item that you want to change. Type your changes, and then tap Save.

   If you want to edit your list on the Amazon app or website, hover over the entry that you want to change, and then hit Edit. Use the text field to type your changes, and then hit

Save. If you are using a mobile device, navigate to Actions -> Edit item.

5. If you want to remove an entry in your lists, pull up the Alexa app, and then tap on the downward-pointing arrow beside the item that you want to remove. Tap the Delete item option. If you want to delete all items in your list, tap View selected, and choose the Delete all option.

   To delete items in the Amazon app or website, click on the checkbox beside the item that you want to remove, then choose Delete Selected. If you wish to delete all entries, click the Select All box, then choose Delete Selected. To delete entries using a mobile device, choose Actions -> Delete Item.

6. If you want to mark a task as completed, pull up the Alexa app, and then tap the checkbox beside the task. If you want to see all items that you have completed, tap the View completed option.

   If you want to do this on the Amazon website instead, click on the checkbox beside the item, and then click the option Mark Selected Complete. If you want to see all the items that you have completed, click on the View completed option. If you are doing this on a mobile device, tap on Actions -> Mark Completed option instead.

Linking Third-Party Services

If you want to access your to-do and shopping lists available in third-party productivity services, you can link Alexa to your third-party accounts using your Alexa app. This will allow you to manage your task lists and remember items that you need to buy more efficiently. Right now, Any.do and Todoist is supported by Alexa.

To link third-party productivity services, pull up your Alexa app and tap on the left navigation panel. Navigate to Settings -> Lists and tap on the Link option on the selected third-party service. Input your username and password for the service, or if you are not registered to any of them yet, create a new account instead. Afterwards, follow the instructions that will appear.

If you wish to unlink Alexa from these services, simply select the Unlink option in the Alexa app.

Note: When you link third-party services to your Alexa account, you are allowing these services to access and make changes to your lists information that you made in Alexa. This may mean that you need to subscribe to terms and policies that the service may require. Also take note that the Alexa app may not be able to support all features that you enjoy in the third-party service.

Adding Your Calendars to Alexa

If you are using the Google Calendar to manage your schedules and events, you can have your voice assistant Alexa access and manage your schedules as well. Once you are able to link your calendar to your Alexa account, you will be able to ask Alexa to read upcoming events or add your schedule to your calendar. If you belong to an Amazon Household, you will be able to add calendars that other account holders have using Alexa as well.

Note: Everyone will be able to hear events that all linked accounts through Alexa have.

To link your Google calendar to Alexa, follow these steps:

1. Pull up your Alexa app and tap the left navigation panel.
2. Navigate to Settings -> Calendar
3. Choose Google Calendar
4. Tap the option Link Google Calendar account.

5. Sign in to your Google account. If you do not have an existing account yet, create an account on https://calendar.google.com.

Tip: Make sure that you have selected the calendar that you want to access before attempting to add any event.

Alexa Commands to Manage Calendars

Here are the commands that you can use in order to review and add events to your Google Calendar.

1. To hear about your next calendar event, say your wake word plus the following:
   - What's on my calendar?
   - When is my next event
2. If you want to hear about an event that is set on a specific day or time, say your wake word plus the following:
   - What's on my calendar on [date]?
   - What's on my calendar tomorrow at [time]?
3. If you wish to add an event to your calendar, say your wake word plus the following:
   - Add [name of event] to my calendar for [date] at [time].
   - Add an event to my calendar. (Alexa will provide further instructions on how you can add an event.)
   -

IFTTT and Alexa

IFTTT, or "If This, Then That," is a service that allows you to be productive by automating your apps, devices and websites by following certain rules, which are also known as applets. When you allow an applet through the IFTT site, Alexa will activate this particular applet when you talk to your Echo Dot.

Here are some examples on how IFTTT works:

- If you ask Alexa to search where your phone is, IFTTT will trigger an applet that will make your phone ring.
- When you command Alexa to tell you your favorite team's upcoming game, an IFTTT applet will trigger and then automatically create a reminder for the game schedule in your Google Calendar.
- When you ask Alexa to read your Shopping list, you will be able to receive your Shopping list in your email.
- When you accomplish a task on a to-do list, you can trigger an IFTTT applet that will tweet your recent accomplishment to all your Twitter followers.

If you wish to use IFTTT, you can create recipes that will essentially create applets for automation. Alternatively, you can also make use of existing user recipes that are available online.

Note: IFTTT also allows you to create actions or applet results. Right now, your Echo Dot cannot be the resulting action of an applet. For example, you cannot have Alexa tell you if you have new tweets available on your Twitter account.

Setting Up IFTTT and Alexa
To setup your IFTTT service on Alexa, go to this link: https://ifttt.com/amazon alexa. This will lead you to the Amazon Alexa Channel on the IFTTT website. Once you are there, follow these steps:

1. Sign in with your IFTTT user credentials. If you do not have an account yet, hit Sign Up and follow the prompted instructions to create an IFTTT account.
2. Once you are signed in, choose the Connect option.
3. Pull up your Amazon account and then link your IFTTT account.

Note: If you are going to use IFTTT with your Echo Dot, remember that it is not an Amazon service – this means that there might be some terms that you need to subscribe to when you are using this service. If you wish to unlink Alexa services from IFTTT, all you need to do is to go to Manage Login with Amazon (https://www.amazon.com/ap/adam).

Use Voicecast to Send Content to Your Fire Tablet

If you own a Fire Tablet, you can use the Alexa service called Voicecast, which will allow you to command Alexa to send more details about a particular topic or information that you are interested in to your Fire Tablet. If you want to have your tablet receive this information automatically, you can turn on Automatic Voicecast.

What happens when you send content using Voicecast to your Fire tablet? The information that you requested will appear right on the lock screen, or if you are using the tablet during the time that you requested Alexa to send over content, you will see a notification on your tablet's Quick Settings menu.

Note: You will not be able to use Voicecast to send any content from your Fire tablet.

What are the services that you can use with Voicecast? You can use this service with these Alexa features:

- Help
- Music
- Questions and answers
- Weather
- Flash Briefing
- Timer and alarms
- Wikipedia

If you wish to activate Voicecast, pull up your Alexa app on your Fire tablet (you will be able to use this service if you have a tablet that has a Fire OS 4.5.1 or later). Afterwards, select Settings from the left navigation panel, and then tap on the Voicecast option. If you wish to receive content on your tablet automatically, also select the Automatic Voicecast (you can find this on the Voicecast menu).

Once you have activated Voicecast, you can make a request to Alexa and then say "Send that to [linked device]" or "Show this on my Fire tablet" once Alexa responds to your request.

Chapter 8: Shopping with Echo Dot

Since you own an Amazon product, you will be able to shop online with ease using your Echo Dot by asking Alexa to place your orders for physical products or digital content over Amazon Prime.

When you ask Alexa to buy something online, your voice assistant will look at the following:

1. Your order history – Alexa will search for Prime items that you have ordered in the past.
2. Amazon's Choice – Alexa will search for items that are priced well, readily available to be shipped, and highly rated by other customers.
3. Prime-eligible items – these items allow you to enjoy Prime now delivery

If the item that you requested is available, Alexa will tell you the product name and its price. If the items that you wish to order are not available to ship using Prime's standard two-day shipping, you will hear the estimated delivery information. Afterwards, Alexa will ask you whether you wish to confirm or cancel your order.

If Alexa is not able to find the item that you want, or if it is not able to complete your purchase, your voice assistant will offer to do one of the following:

1. Add the item/s to your Amazon cart
2. Add the item/s to your Shopping List
3. Look at your Alexa app for you to see more options.

Take note that you can also ask Alexa to cancel your orders immediately or track orders that has already shipped.

What You Need to Do to Order
When you place an order through Alexa, your voice assistant will order your items using the 1-Click payment setup that you have

in your account. All the physical items that you have purchased can also be returned for free.

To be able to purchase using Alexa, you will need to have the following:

1. To order digital content from Amazon's Digital Music Store:
   - 30-day free or a paid annual membership for Amazon Prime
   - A payment method that has a billing address in the US and a US bank account in 1-Click Settings
   - A US account for Amazon
2. To order any physical items from Amazon
   - 30-day free or a paid annual membership for Amazon Prime
   - A payment method that has a billing address in the US and a US bank account in 1-Click Settings
   - A US account for Amazon
   - A payment method that has a billing address in the US and a US bank account in 1-Click Settings
3. To track your order's shipping status or to put product in your Amazon shopping cart:
   - An Amazon account

Ordering Music with your Echo Dot

If you want to save a music track and buy the song, you can command Alexa to purchase the song using your 1-Click payment method from the Amazon Digital Music Store. If you have a Prim membership, you can add any Prime Music audio from the Digital Music store to your library for free. Take note that with your Prime membership, you can store any track from the Digital Music Store without having to worry about your storage limit or having to pay for anything.

To purchase music using Alexa, all that you need to have is a US billing address, along with a payment method that uses any US bank or a gift card from US Amazon.com.

Pro Tip: If you want to enter a confirmation code before you place any order or to disable voice purchasing, you can set up these options by pulling up your Alexa app and then navigate to Settings -> Voice purchasing.

Alexa Commands When Shopping for Music

Here are the commands that you can use to buy audio content using Alexa:

1. To buy a song or an album, say your wake word plus the following:
   - Shop for the album [name of album].
   - Shop for the song [track title]
2. To buy songs by a particular artist, say your wake word plus the following:
   - Shop for new songs by [name of artist]
   - Shop for songs by [name of artist]
3. To add or purchase the track sample that you are playing, or the song that you are listening to on an Alexa-supported station, say your wake word plus the following:
   - Add [track/album title] to my library.
   - Buy this [track/album title].

Note: Alexa will notify you about your purchase even if the item that you want to add to your library does not come with any cost.

Shopping for Prime Items

If you have an Amazon Prime membership, you will be able to choose from thousands of Prime-eligible products that are either fulfilled or sold by Amazon. You can also use voice shopping to re-order items you have previously ordered using your Prime account.

About Prime Now Fulfillment

There are some locations in the US that allows you to enjoy two-hour deliveries for free via Amazon Prime Now. By default, Alexa checks if the items that you have ordered can be fulfilled immediately through this service. If you do not want to receive your orders using this expedited option, you can decline this option and look at your Alexa app for more convenient shipping options.

Take note that you cannot specifically request for your orders to be fulfilled by Prime Now at this time. If you are ordering alcohol or items from third-party stores, you will not be eligible to get these orders using this expedited shipping option. Amazon is also not capable of applying any promo credits for any orders fulfilled by this option.

If you have purchased items that are eligible to be fulfilled via Prime Now, you can track your shipment using the Prime Now application or by going to www.primenow.amazon.com. Once your order arrives, you will be receiving a phone call (if you have required a phone call) that you have placed together with your default address in your 1-Click settings. You do not need to be present in order to receive your order.

You are not charged for a delivery tip when you have a Prime Now order, but you can provide a tip up to 48 hours after receiving your orders via the Prime Now app.

Pro Tip: There are eligible locations that will allow you to order meals using Alexa through Amazon. When ordering meals, Alexa will use your default payment and shipping information. You can track your orders from restaurants using the Prime Now app or by logging in at

https://primenow.amazon.com/restaurants/purchases.

Here are the commands that you can use when using Alexa to shop for Prime-eligible items:

1. To buy Prime items (you can purchase up to 12 of any Prime-eligible product), say the wake word plus the following:
    - "Order a [product name]."
    Say Yes or No when Alexa locates the item and then requests that you confirm your purchase.
2. To reorder a product you have previously brought from Amazon Prime, say the wake word plus the following:
    - "Reorder [product name]."
    Say Yes or No when Alexa locates the item and then requests that you confirm your purchase.
3. If you wish to add an item to your Amazon shopping cart, say the wake word plus the following:
    - "Add [product name] to my cart."
4. If you wish to cancel an item right after you placed it, say the wake word plus the following:
    - "Cancel my order."
    Note: You can also cancel any order delivered via Prime Now using your Prime Now app or by going to www.primenow.amazon.com.

Tips:

- If Alexa cannot cancel your order, go to https://www.amazon.com/your-orders to withdraw your purchase. You can manage your restaurant orders through the Prime Now app or by going to www.primenow.amazon.com.

How to Track for Open Orders

If you have placed more than one open order, you can command Alexa to provide you your order status for the one that will be arriving soon.

To track your order status or your recently shipped out order, say the wake word plus the following:

- "Track my order."
- "Where is my stuff?"

Tip: You will be receiving an email or SMS notification that includes your tracking information when you place an order that is going to be fulfilled by Prime Now or Amazon Restaurants.

Managing Your Settings for Voice Purchasing

Once you have registered your Echo Dot, you will be able to perform voice purchasing by default for both digital and physical items for purchasing by default. If you wish to change your voice purchasing settings, pull up your Alexa app and then navigate to Settings -> Voice Purchasing.

To toggle voice purchasing on or off, simple tap the option Purchase by voice. You can also opt to require a 4-digit confirmation code (this will not appear in your voice history) before you complete any purchase that you placed using Alexa by simply entering your desired code and then tapping Save Changes. If you wish to make changes to your 1-Click Settings, tap the option Go to Amazon.com to update your billing address or desired payment method.

Chapter 9: Managing a Smart Home Using Echo Dot

One of the strengths of the Amazon Echo Dot is its ability to control smart home devices – if you have Alexa-compatible light bulbs, speakers, TVs, and other appliances, you can enjoy a household that you can fully control using your voice.

Connecting Your Smart Home Device with Echo Dot

Once you are able to verify that your device is compatible with Alexa, you will be able to use Alexa's skills to make use of your device's features. All you need to do is to download your smart home device's companion app from the manufacturer's website and connect it on the same internet connection that your Echo Dot is connected to.

To connect your device with Alexa, pull up your Alexa app and follow these steps:

1. Select the option Smart Home from the Menu.
2. Tap Get More Smart Home Skills.
3. Search for the keywords that will match the skills that you want. Once you find the skill, select the option Enable Skill.
   Tip: If you cannot find a skill that is applicable to your smart device, your gadget may not be compatible with Alexa.

If you want to remove a device from your Echo Dot, follow these steps:

1. Pull up your Alexa app, and then tap Smart Home from the left navigation panel.

2. Go to Devices and tap the Forget option for the device that you want to remove. You can have your Echo dot forget all the devices that you have connected by disabling the skill that they perform. You can find this option under the Skill section in your app.

Note: Turning off your Echo Dot will not remove the device that you have already connected. Your Echo Dot will still be able to recognize all devices that you have linked with it once you plug it in.

Smart Home Gadgets Supported by Alexa

Before you connect a smart device to Alexa, the first thing that you need to do is to check whether it is compatible with your Echo Dot or not. Here is the list of all the devices that are compatible and optimized to be used with Alexa:

1. Lighting Equipment
   a. Stack Lighting Solutions
   b. TP-Link Smart LED Light Bulbs
   c. iDevices Socket
   d. TP-Link Smart Wi-Fi Switch
   e. Caseta Wireless Dimmer Kit w/ Smart Bridge
   f. Caseta Wireless Plug-In Lamp Dimmer Kit w/ Smart Bridge
   g. WeMo Light Switch

   h. LIFX White A19 Smart LED Light Bulb
   i. Philips Hue Lighting Products
   j. GE Link Smart LED Light Bulbs
   k. LIFX Color BR30 Smart LED Light Bulb
   l. Caseta Wireless In-Wall Dimmer Kit w/ Smart Bridge
3. Switches and Outlets

a. WeMo Electronics Insight Switch
b. TP-LINK HS110 Wi-Fi Smart Plug
c. iDevices Outdoor Wi-Fi Switch
d. D-Link Wi-Fi Smart Plug
e. WeMo Electronics Switch
f. TP-LINK HS100 Wi-Fi Smart Plug
g. iDevices Wi-Fi Switch
h. iHome Wi-Fi Smart Plug

4. Thermostats and Other Devices
   a. ecobee3 Smarter Wi-Fi Thermostat
   b. Tado Smart Temperature Control Programmable Air Conditioner
   c. iDevices iShower2 Bluetooth Speaker
   d. Honeywell Lyric Thermostat
   e. Venstar T7900 Colortouch Wi-Fi Thermostat
   f. Haiku Home L Series Smart Ceiling Fan
   g. Nest Learning Thermostat
   h. Sensi Wi-Fi Programmable Thermostat
   i. iDevices Thermostat
   j. Honeywell Smart Thermostat
   k. Carrier Cor 7-Day Thermostat
   l. Venstar Residential Voyager Thermostat

How to Command Alexa to Control Smart Home Devices

Once you are able to confirm that your device is compatible with Alexa, you can start enabling your voice assistant to manage your smart home device.

Note: There are some skills that require you to say "Open [name of skill]" before you can make Alexa perform your request.

Here are the commands that you can use to operate your smart home devices using Alexa:

1. If you wish to turn on or off your device, say the wake word plus the following:
   - "Turn on / off [group name/name of smart home device]"

2. To activate or deactivate a scene, say the wake word plus the following:
   - "Turn on / off [group or scene name]"
3. To control the brightness of Alexa-compatible lights, say the wake word plus the following:
   - "Brighten / dim [group or smart home device name]"
   - Set [group or smart home device name] to [brightness percentage] percent"
4. To set the temperature of your thermostat, say the wake word plus the following:
   - "[Increase or decrease] the [group or smart home device name] temperature"
   - "Set [group or smart home device name] temperature to [desired temperature] degrees"
5. To set your electric fan's speed, say the wake word plus the following:
   - "Set my bedroom fan to [desired speed] percent."
6. To activate an activity or a channel, say the wake word plus the following:
   - "Turn on [activity or channel]"

Notes:

- There are some skills that may support different features of the device that you wish to control. To get more information about your device's actions, pull up your Alexa app and go to the smart home skill.
- You can see available scene names in the Alexa app by going to Smart Home -> Your Devices.
- You will be able to make and manage scenes for your devices using their corresponding companion app.

Great Alexa Skills for Your Smart Home Devices

Your Echo Dot already comes with its own smart home skills, and it pays to know what additional skills you should teach to Alexa based on the smart devices that you have in your household.

If you are interested in using Alexa mainly for smart home control, you can find some skills in the internet of things (a development that allows smart appliances and other objects that uses the internet to get and send data), but there are only a few things there that may prove to be useful. This list will be extremely helpful in knowing which skills you should activate in your Alexa app based on the devices that you own or considering to buy:

1. Logitech Harmony
    This skill allows you to use Alexa to have total control in turning your cable box, TV, and the AV receiver. It also allows you to make use of customized names which will be extremely handy when you wish to activate certain entertainment services, such as Netflix. You also get to enjoy a nice, streamlined voice control when you set this skill up.

    This remote companion allows you to turn your TV and cable on or off without having to rely on any IFTTT trigger or other devices or skill. However, the downside is that it does not allow you to tweak the volume, pause, or play your show using your voice.
2. Smart Things
    This skill, which is developed by Samsung, allows you to control your home's lighting more efficiently and turn smart appliances such as heaters and electric fans on or off. You will need to purchase the Smart Things hub first and see to it that your smart devices can connect to it in order to use this skill.
3. Wink

If you do not want to get the Smart Things hub, you can get the less expensive Wink hub instead. This skill also works with thermostats from Nest and is incredibly easy to set up.

4. Nest

This skill is a must have if you have a Nest thermostat that is not connected to any hub – you can set, increase, or decrease the temperature to your desired degree. The only downside is that it is not yet capable of alternating from cooling and heating, and that Alexa will not be able to tell you what the current temperature is.

5. Insteon

This skill is dedicated to the Insteon starter kit, which is considered to be the less popular brand that serves as a great alternative to Smart Things and Wink. This starter kit is known for providing you a variety of items for a lower price, which includes two modules for dimming your house lights and turning off your appliances.

6. TP-Link Kasa

If you are just starting in using smart home devices, it will be helpful to activate the TP-Link Kasa skill. This allows you to control your smart home devices without having to rely on a hub – instead on having to buy a starter kit or a hub, you will be able to use a plug instead.

7. Caseta Wireless

This skill is considered to be one of the best light dimmers out there – if you have the Caseta Wireless dimmer kit, you will also be able to use a remote that you can use to dim, turn on or off a light during times that you cannot use your voice. The only downside is that it is not optimized to be used with scenes.

8. Vivint

Vivint is actually a third-party skill, but it is very effective as a premiere skill for your smart home. It provides you control over your entire smart household, which includes thermostats, security, and lighting. Take note that having Vivint as your main control is something that is not recommended for beginning smart household owners – it entails a lengthy setup and having an extensive grasp in smart home technology.

9. LIFX

This skill is to be used with the LIFX LED lights, which allows you to change the color of your light bulbs and dim them as well. This means that you can change your bulb to display different shades of white light. To change colors, say "Alexa, tell 'Life-Ex' to make my lights [your color choice]."

10. Haiku

This skill allows you to control your ceiling fans with your voice without having to depend on a smart switch or plug. The setup and control is also seamless.

Chapter 10: Other Skills You May Want to Try Out

Right out of the box, your Echo Dot can already do many things that are designed to make your life so much easier, such as integrating with IFTTT recipes, ordering millions of items from Amazon, and streaming music from supported music services. What makes your investment in your Echo Dot worth it is that the number of skills that Alexa can perform is continuously growing, thanks to the numerous brands and developers that are creating apps that make use of Alexa.

Finding Skills

There are more than 4000 skills (as of November 2016) that Alexa is able to perform, and because of the huge number of skills that are available, you will definitely not want to just randomly add features to your Echo Dot. What you will want to do is to discover new skills and add the ones that you are interested in.

Amazon has developed a skill named Skill Finder that will allow you to find new skills. To launch this, you will need to say the wake word and then "tell Skill Finder to give me the skill of the day." Alternatively, you can also say the wake word and then "open Skill Finder."

Recommended Skills

Here are some of the most useful Alexa skills that you may want to activate:

1. Capital One

This skill lets you view credit card bills and place your payment for the ones that are due. This service is very secure – it will prompt you to log in using your account details and also perform a security check. When you want to use the skill, you will be prompted to provide a four digit code to verify your identity. All you need to do is to take caution when you are using this skill in the presence of other people – anyone that has overheard you saying your personal code will be able to access your saved banking information and your credit card by requesting it from Alexa.

2. Giant Spoon

If you are in the field of marketing and you are brainstorming for creative ideas, this skill will be extremely handy. Of course, there is a chance that the idea that you will get may not be very useful to the project that you have, but you will definitely get interesting concepts that you may be able to use for other things.

3. Opening Bell
This skill checks stock prices, which will be very handy if you are planning to invest in the stock market. Using this skill, you do not need to worry about the stock or ticker symbol – you can check for prices by using the company's natural name. For example, if you want to check for Facebook's stock price, you can simply say "Alexa, ask Opening Bell for the price of Facebook."

4. Yonomi
Yonomi functions similarly to IFTTT, but for smart home devices. This skill will enable you to create a virtual device for scenes that you were able to create, which makes your command become natural-sounding.

5. Automatic
Automatic is a small device that you can attach to the OBDII port of your car, which will allow you to track your automobile's status using your smart phone. The

Automatic skill will allow you to ask Alexa to ask for your car's mileage or fuel level.

6. MySomm

   MySomm is an excellent skill for foodies that will want to know what kind of wine pairs with a particular food. To do that, you can say this command:

   "Alexa, ask Wine Gal what goes with [type of food or name of dish]."

7. The Bartender

   If you are the type of drinker that wants to know what goes into your drink or if you want to recreate a cocktail in your own home, this is the skill for you. This skill allows you to hear your favorite drink's ingredients and recipe. Since The Bartender has a robust library, it is very likely that you will get the recipe for virtually every drink that you have already encountered.

8. Meat Thermometer

   This skill is great for anyone who likes to cook their own food and would like to know internal temperatures that are safe for various types of meat while they are being cooked. For example, you can say use the following command:

   "Alexa, ask Meat Thermometer what is the best temperature for salmon."

9. Domino's

   This skill allows you to access Domino's Easy Order by simply saying this command:

   "Alexa, open Domino's and place my Easy Order."

   This skill also allows you to track your order status by saying the following command:

   "Alexa, open Domino's to track my order."

10. Kayak

    This skill allows you to do some research about your trips and vacations. If you wish to know how much you need to save up for fare or know your travel options, this skill will prove to be extra helpful.

    For example, you can use the following commands for your travel research:

"Alexa, ask Kayak where I can go for $50"
"Alexa, ask Kayak how much it costs to fly from Queensland to California."

11. Uber

If you want to get a car to get to anywhere, you can say the following command:

"Alexa, ask Uber to get me a car."

12. Lyft

Lyft is a service that is very similar to Uber, except that you can know how much you will need to pay before your order your ride. For example, you can say the following:

"Alexa, ask Lyft how much a Lyft from home to the gym costs."

13. Ask My Buddy

This skill functions as your personal alert system. If you need help, you can ask Alexa to send an SMS to a friend or your relatives, or even place a call when needed. Take note that this skill will not call 911, but you can reach anyone that is in your contact list in times of emergency.

14. Jeopardy!

This skill allows you to enjoy your daily dose of trivia in the style of the game Jeopardy. Just like in the show, you will need to respond to Alexa by providing an answer to the question, but you only have a few seconds to do so. You are going to hear 6 questions from different categories.

15. The Magic Door

This skill allows you to play a game where your choices affects what happens next in your adventure, which includes characters like sleeping dragons, temples, hills, and forests. It's not clear what happens next in your adventure, but as long as you are playing the game, more options will unravel. If you have kids, this skill will be a good addition to your household. Take note though that the game does not have a stop and start option, which means that if you stop using the skill, the adventure you have created also ends.

16. Short Bedtime Story

This is another skill that is worth teaching to Alexa if you have kids – what you're getting out of this skill is a 30-

seconder story that is customized to feature the name of your child.

17. Fitbit

    Fitbit rivals Alexa for being one of the most downloaded free apps. When you use the Fitbit skill, you will be able to use your Echo Dot to tell you information about your physical activities and your health's status, such as the amount of sleep you were able to get the previous night and your heart rate while you are at rest. The integration of Fitbit with your Echo Dot allows you to enjoy fast updates on different stats that you may want to check on a regular basis, which include exercise goals and the number of steps you have taken.

18. The 7-Minute Workout

    This skill is considered as one of the most popular skills available in the Alexa app. Just like what its name says, this skill is designed to make you perform quick yet fat-burning workouts while tracking your progress in routines that you are doing. If you are not able to complete a routine, you can pause it and then resume it for a later time.

19. Campbell's Kitchen

    This skill serves as a personal cookbook that contains a huge amount of recipes based on main ingredients that you have or the type of dish that you want to cook. You will be able to also get the full recipe in your email, just in any case you will still want to pull up a written recipe that you can follow.

20. 1-800 Flowers

    As the name of the skill suggests, this skill allows you to send flowers to your loved ones using only your voice. This skill also provides you the ability to select which flower options work best for the occasion or the receiver, as well as the day of the delivery.

Chapter 11: Troubleshooting and Optimizing Your Echo Dot

While your Echo Dot and Alexa are designed to make things easier for you, there are times wherein not all things will work out fine while you are using them. This chapter provides you answers to most common Echo Dot and Alexa problems encountered, as well as some tips and hacks that you can use to ensure seamless use of your device.

Your Device Can't Connect to Wi-Fi

If your Echo Dot is not able to connect to your Wi-Fi network at home, make sure that you are connecting to a dual band Wi-Fi (5GHz or 2.4 GHz) that makes use of the 802.11 a/b/g/n standard. It will not be able to connect to any peer-to-peer or ad-hoc network.

You will also be able to check the current status of your network by looking at Echo Dot's power LED, which can be found close to the adapter port. If you see that the power LED light is solid white, it means that your device is connected to Wi-Fi. If it displays a solid orange light, your Echo Dot does not have internet connection. If you see that it is blinking, your device is connected to Wi-Fi, but is unable to connect to the Alexa Voice service.

If you are having trouble connecting your Echo Dot to your home's Wi-Fi, try this things first:

- Make sure that you have entered your Wi-Fi network's password correctly.
- See to it that you have the latest firmware for your router or modem.

- If you have your Wi-Fi password saved to Amazon but you changed your connection's password recently, see to it that you reenter the new password to be able to connect to your network.
- If your router is using WPA+WPA2 encryption for its password, try changing the encryption type to WPA2 or WPA only. If you can change the encryption type, try setting it to AES.

Once you are able to perform these things, attempt to connect to your Wi-Fi again.

Reduce Congestion

If you have different devices connected to a single Wi-Fi network, it is possible that you are unable to connect your Echo Dot to the network due to congestion. Try doing the following:

- Disconnect wireless devices that you are not using
- Move your Echo Dot closer to the modem or router if your device is in another room
- Make sure that the device is not near a baby monitor, microwave, oven, or any device that may cause signal interference
- If available, use the 5 GHz band for your router. This frequency is often less congested and has better range.

Alexa Cannot Understand You

There will be times wherein Alexa does not seem to respond to your commands or requests because it is unable to understand what you are saying. To prevent this, place your device at least eight inches away from any item that may cause interference or walls. If your device is on the floor, place it on a table or a higher location. Turn off music and remove excessive background noise if there is any.

See to it that you are specific with your request or question. You can try to rephrase the question to make it seem less general. You may also want to check with your Alexa app to find out what Alexa heard and send the appropriate feedback to Amazon (tap the More button on the interaction card, which is located on the Home screen)

To make Alexa learn how to understand you better, it is recommended that you use Voice Training at least once a day.

Your Device Doesn't Respond or Turn On

If your device is not turning on or responding to any of your requests, follow these troubleshooting steps:

- Check if you have plugged your Echo Dot using the power adapter that is included in the box. If you are using a cell phone charger to power your device, it may not be providing enough power for your Echo Dot to turn on or perform correctly.
- See to it that you have placed your device at least three feet away from a speaker that it is connected to. When Echo Dot is too near to an external speaker, it may have difficulty hearing the wake word and any command.
- Make sure that there is no background noise while you speak.
- Try repeating your commands and say them clearly and naturally.
- Move your device at least eight inches from any walls or any object that may cause interference.

Your Alexa App is Not Working Right

If you can't open the Alexa app or if you keep on receiving error messages, here are some solutions that you may want to try:

Check if the device that you are using is compatible with the app.

The Alexa app only works with devices that has the following:

- Android 4.4 and up
- iOS 7 .0 and up
- Fire OS 2.0 and up
- Web browsers Safari, Firefox, Internet Explorer 10, Microsoft Edge, and Chrome.

Troubleshoot Alexa App with your devices by

- Restart your iPad, iPhone or iPod touch.
- Force close the Alexa app.
- Uninstall and reinstall the app.
- If you are using a web browser, try reloading the web page. If it still doesn't work, try deleting the cookies and clearing the cache, then restart your browser.
-

Echo Dot Does not Connect to Bluetooth

If you are having a hard time connecting your mobile device (i.e. tablet or smart phone) to your Echo Dot, see to it that your device has the following Bluetooth profiles:

- A2DP SNK (Advanced Audio Distribution Profile)

  This profile enables you to stream audio from a connected mobile device, such as a tablet or a phone, to your Echo Dot. Take note that Echo Dot is also capable of supporting the A2DP SRC profile, which enables the device to stream audio to a Bluetooth speaker.

- AVCRP (Audio/Video Remote Control Profile)
  This profile enables you to control your devices by using only your voice when they are connected to your Echo Dot.

Check your Bluetooth device's batteries

One of the reasons why your Bluetooth device cannot connect to Echo Dot is that it probably does not have enough power to connect wirelessly. If your device uses a non-removable battery, check if your device is charged. Try using new batteries if your device uses removable ones.

Look out for interference

If your Bluetooth device or your Echo Dot is too close to interference sources, such as baby monitors, ovens, and other wireless devices, you may be experiencing connectivity problems due to interference. Move them away from such devices, and then try moving your Bluetooth device closer to your Echo Dot when you pair it.

Try pairing your Bluetooth device again.

If your device still does not connect to Echo Dot, you can try clearing all the devices connected, and then attempt to connect them back again. To do this, pull up the Alexa app, and then tap Settings on the left navigation panel. Choose Echo Dot from the list, and then hit Bluetooth. Select the Forget option next to all Bluetooth devices. Once you cleared all paired devices, restart your Echo Dot and your device. Turn on the Bluetooth on your mobile device and then pull up your Alexa app. Pair both devices by navigating to Bluetooth -> Pair a New Device. Once Alexa enters the pairing mode, select Echo Dot on the Bluetooth settings of your mobile device. You will get a notification from Alexa once the devices are paired successfully.

Alexa Can't Find Your Smart Home Device

If you can't connect your smart home device with your Echo Dot, follow these troubleshooting steps:

- Check if your device is compatible with Alexa. You can check Chapter 9 for a list of compatible devices or go to https://www.amazon.com/alexasmarthome.
Note: You can also check the Skills on your Alexa app to get more information about devices that can work with your Echo Dot.
- Get the companion app that goes with your smart device to set it up with Echo Dot. You will usually find this app on the manufacturer's website.
- Try restarting Echo Dot and the smart device.
- Try disabling and enabling the smart home skill in your Alexa app.
- Check for any software or firmware updates for your smart devices.

Discover Devices without Skills

There are some devices (such as Wemo and Philips) that do not require an Alexa skill to be discovered. Here are some tips on how to discover smart home devices without using a skill:

For Philips smart home devices:

- Press the button that you will find on the device's bridge
- Say the wake word, and then "Discover devices."

For Wemo smart home devices:

- Say the wake word, and then "Discover devices."

Use the same network for Echo Dot and your devices

You will need to connect your smart home devices to the same Wi-Fi network that you are using with Echo Dot. Also take note that the Echo Dot and smart home devices work fine with personal networks, but work or school networks may prevent you from connecting these devices to them.

It will also help to check if your smart home devices can connect to a dual band Wi-Fi – there are some devices that are only capable of connecting to the 2.4 GHz Wi-Fi. If your Echo Dot is connected to your Wi-Fi network using the 5 GHz frequency, you might want to change your band connection and then update the Wi-Fi settings for Echo Dot using the Alexa app.

Check the name of your smart home device group

See to it that you are using a group name that is easily understandable by Alexa when you say the group's name out loud. For example, if you have named a group using your mobile device to be "b3droom lights", change it instead to "Bedroom Lights".

Hacking Alexa

There will be times wherein you will want your Echo Dot and Alexa to perform actions and modify commands and skills that it already knows. This hack will enable you to make use of a source code to modify, redirect, or create an action for some triggers.

Take note that hacking Alexa may require intermediate to advanced knowledge of computer networking, but the steps that will be outlined below are relatively easy to follow. If you are unaware of some terms used here, you can easily look up what they mean online.

1. Perform a port scan on your Echo Dot using the setup and the regular mode. Keep in mind that Alexa creates the Amazon DP-1, which is also known as an infrastructure Wi-Fi network. A port scan is a network scan that checks all the IPs that are active on a network, as well as the TCP

and UDP ports that are in use. To perform a port scan, you can use scanners such as Wireshark and NMAP.

You can try scanning the address 192.168.11.44. When you go to the DHCP, you will find that you are automatically rerouted to the address 192.168.11.1. You will notice that you are getting responses from the ports 8000 and 443.

2. After the scan, you will see that Alexa runs on mDNS. The SSH will appear to be running on port 22 along with the SFTP. Workstation will be running on port 9.

3. Now, use a Wireshack on Alexa. Keep in mind that Alexa typically works on two pins online, which are pindorama.amazon.com and pins.amazon.com.

4. At this point, you can now start changing Alexa's configuration process. To do this, connect to the DP-1 network by configuring the HTTP POST entry to 8080. Afterwards, go to /OOBE HTTP/1.1, and then enter Thrift JSON on the text field. You will be able to access the [0, 1:{"str","ping"}, 0, {}] configuration port.

5. Once you are on the configuration port, you will be able to see all source codes that you can use as they are. Now, you will be able to tweak different commands that you can use on your Echo Dot by working on the following APIs:
   a. prod: "projectabc.amazon.com"
   b. beta: "projectabc-ui-registration.aka.amazon.com"
   c. alpha:"projectabc-ui-dev-registration.aka.amazon.com"
   d. gamma:"projectabc-ui-gamma-registration.aka.amazon.com"

Using Alexa with Unsupported Smart Devices

Alexa is supposed to only be compatible with devices that are in their list of supported devices. However, there is a way for you to connect certain smart devices with your Echo Dot and operate them with your voice.

For this hack, you will discover how to create a hack that will allow you to connect smart devices to Echo Dot using the UPnP, or Universal Plug and Play. This protocol is used by any network that is capable of discovering, collaborating, and configuring any device that requires being connected through a network. This protocol is used by tablets, smart phones, computers, scanners, gaming consoles, and so on. This means that you can actually connect Alexa devices such as the Echo Dot to devices that makes use of this protocol. For example, you can actually have your Echo Dot connect to a printer and request for a print job using the UPnP.

Note: Doing this Alexa app will require you to make use of a programming language and a text editor.

For this example, you will be able to connect different fake WEMO devices to connect to Alexa using the UPnP.

To find devices and connect them to Alexa

- Perform a UPnP broadcast over UDP, and then perform MSearch for URN Belkin Device – this will make your Echo Dot search for your WEMO devices.
- Perform an HTTP over UDP over http://<ip>:49153/setup/.xml – Your WEMO device will start to respond with URL.
- Perform HTTP over TCP through Get/setup.xml – this will prompt your Echo Dot to request for a description of your WEMO device.
- Perform HTTP response – Your WEMO device will return with the description of device.

When you are able to perform all these, you will be also be able to create your own WEMO cloud. This cloud will serve as the base for Echo Dot to receive instructions and have Alexa understand what it is supposed to do. Afterwards, you will be able to provide commands to your cloud that your Echo Dot will be able to understand. Here is what you need to do to create your cloud for

WEMO devices, and virtually all other devices that can connect via UPnP:

1. Create an IP address for each virtual switch that want to create. If you want a separate switch for your light bulbs, kitchen appliances, and thermostat, you will probably need to have three addresses.
2. Create a listener for broadcasts using UDP and link it to the address 239.255.255.250. You can find this on port 1900. Essentially, this address and port is the one that is used by the UPnP, which will enable Alexa to find your devices.
3. Create a listener on the port 49153 for all of the IP addresses that you created earlier.
4. Download setup.xml. This document will allow you to tweak it in such a way that it conforms to the protocol – doing so will allow Alexa to receive the right data from your devices.
5. At this point, Alexa will be able to send you commands that will enable you to turn on or off your devices. You will be able to change these commands to have Alexa perform actions that you want for your devices and not just be limited by turning them on or off. For example, you can ask Alexa to change the thermostat thermometer at a specific part of the day.
6. Now, go to the address 239.255.255.250:1900 on port 5000 in order to perform a search on Alexa. To do this, type the following request:

```
M-SEARCH * HTTP/1.1
HOST: 239.255.255.250:1900
MAN: "ssdp:discover"
MX: 15
ST: urn:WEMO:device:**
```

7. If you do not see a TCP on your screen, then you get the idea that Alexa was not able to make a connection. This will mean that if you issue one command to Alexa, you will

be getting too many responses from it and it is possible that Alexa will be performing an action that you don't want it to do. For this reason, you will need to keep on tweaking ports by using this code:

```
HTTP/1.1 200 OK
CACHE-CONTROL: max-age=86400
DATE: Mon, 12 Oct 2015 17:24:01 GMT
EXT:
LOCATION: http://192.168.5.190:49153/setup.xml
OPT: "http://schemas.upnp.org/upnp/1/0/"; ns=01
01-NLS: 905bfa3c-1dd2-11b2-8928-fd8aebaf491c
SERVER: Unspecified, UPnP/1.0, Unspecified
X-User-Agent: abcred
ST: urn:WEMO:device:**
USN: uuid:Socket-1_0-221517K0101769::urn:Belkin:device:**
```

8. See to it that the port changes that you have made make your devices perform actions that you expect them to do. You can do this by creating changes on socket 1-0 by using the following code:

    ```
    GET /setup.xml HTTP/1.1
    Host: 192.168.5.189:49153
    Accept: */*
    ```

9. Once you are able to make those changes, the switch that you have created will return with the following result:

```xml
<?xml version="1.0"?>
<root xmlns="urn:Belkin:device-1-0">
<specVersion>
<major>1</major>
<minor>0</minor>
</specVersion>
<device>
<deviceType>urn:Belkin:device:controllee:1</deviceType>
<friendlyName>kitchen light</friendlyName>
<manufacturer>Belkin International Inc.</manufacturer>

<manufacturerURL>http://www.belkin.com</manufacturerURL>
>
<modelDescription>WEMO Plugin Socket
1.0</modelDescription>
<modelName>Socket</modelName>
<modelNumber>1.0</modelNumber>
<modelURL>http://www.belkin.com/plugin/</modelURL>
<serialNumber>221517K0101769</serialNumber>
<UDN>uuid:Socket-1_0-221517K0101769</UDN>
<UPC>123456789</UPC>
<macAddress>94103E3489Co</macAddress>
<firmwareVersion>WeMo_WW_2.00.8326.PVT-OWRT-
SNS</firmwareVersion>
<iconVersion>0|49153</iconVersion>
```

10. You will see that the Friendly Name indicated in the above code is the action that will Alexa will perform when you turn a device on or off. The Serial Number UDN will decipher the data that is created by the device's mechanism. For this to work, you will need to create an HTTP response, which will look like this:

```
POST /upnp/control/basicevent1 HTTP/1.1
Host: 192.168.5.189:49153
Accept: */*
Content-type: text/xml; charset="utf-8"
SOAPACTION: "urn:Belkin:service:basicevent:1#SetBinaryState"
Content-Length: 299

<?xml            version="1.0"            encoding="utf-8"?><s:Envelope
xmlns:s="http://schemas.xmlsoap.org/soap/envelope/"
s:encodingStyle="http://schemas.xmlsoap.org/soap/encoding/"><s:Bod
y><u:SetBinaryState
xmlns:u="urn:Belkin:service:basicevent:1"><BinaryState>1</BinarySta
te></u:SetBinaryState></s:Body></s:Envelope>
```

Now that you have completed your code, you will be able to see it on your screen. Alternatively, you can also create a similar yet simpler code using that programming language Python.

Conclusion

Thank you for reading this book! I hope that this book served as a guide on how you can use your Echo Dot and optimize it for your needs. I also hope that this book helped you discover Alexa's skills and perform actions that will help you manage your day-to-day activities by using your voice assistant as an essential addition to your smart phone applications and devices.

The next step is to discover other Alexa skills that will help you maximize your Echo Dot – you can do that by checking your Alexa app from time to time about new skills that are added by different users and brands that integrate their products with Alexa. Alternatively, you can also start exploring Alexa's Developer API and the Smart Home Skill API to create skills that are customized for your needs. To do this, all you need to do is go to https://developer.amazon.com/alexa-skills-kit to start learning how to build customized skills.

Finally, if you enjoyed reading this book and found it useful in using your Echo Dot the way you want it to perform, please take the time to leave a comment and a rating for this book on Amazon.com. I'm looking forward to hearing from you soon!

Printed in Poland
by Amazon Fulfillment
Poland Sp. z o.o., Wrocław